yes, we can! She Can STEM

BIOLOGY & ECOLOGY

QUARRY

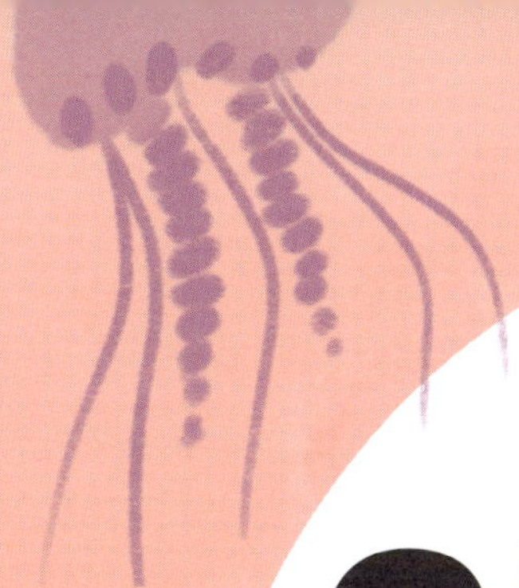

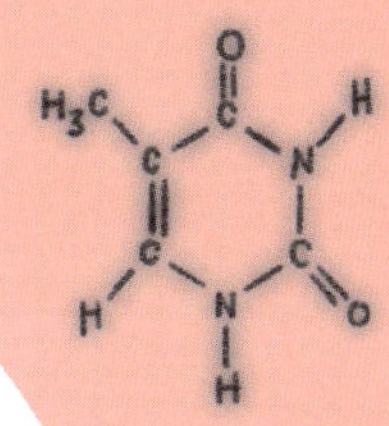

yes, we can!

She Can STEM

BIOLOGY & ECOLOGY

Inspiring Stories of Women Who Changed the World

LIZ LEE HEINECKE

The Kitchen Pantry Scientist

QUARRY

Quarto.com

This library edition published in 2026 by Quarry Books,
an imprint of The Quarto Group.

100 Cummings Center, Suite 265-D, Beverly, MA 01915, USA.
T (978) 282-9590 F (978) 283-2742

Distributed in the United States and Canada
by Lerner Publisher Services
241 First Avenue North
Minneapolis, MN 55401 U.S.A.
www.lernerbooks.com

First Library Edition

Library of Congress Cataloging-in-Publication Control Number: 2025032029

Printed in China

10 9 8 7 6 5 4 3 2 1

To each woman, famous or unseen,
who paved the way for women in science.

And to those treading the road today
to make a better future for all of us.

CONTENTS

Biology

Ecology

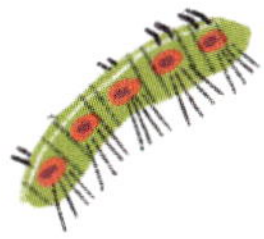

INTRODUCTION

When you think of a scientist, what image pops up in your mind? Do you picture an old man with crazy hair? That's how the world has been portraying scientists for a long time. Luckily, the face of science has changed dramatically and is continuing to evolve. Today, everyone should be able to see themselves as a scientist.

Women have been involved in STEM (Science, Technology, Engineering, and Math) for thousands of years and today, there are more women than ever working in these disciplines. This book series lays out the biographies of 50 amazing women who are part of the scientific landscape, from historical scientists to modern ones. Many are role models who have inspired other girls and women to chase scientific dreams.

For example, in 1898, a physicist named Marie Curie became interested in invisible energy given off by some rocks. She figured out how to measure the energy and named it "radioactivity." Soon, she and her husband Pierre discovered two new radioactive elements and named them radium (Ra) and polonium (Po). Marie then invented a method to purify radium and won two Nobel prizes for her work, one in physics and one in chemistry. Few females did science at that time, but when girls and young women read about Marie's work, they realized that they could be scientists too.

Another scientist featured in this series helps us imagine how it might have felt to be the only woman in a large physics class in the 1960s. Day after day, Jocelyn Bell Burnell was harassed by fellow university students who stomped their feet, whistled, and banged on their desks as she walked through the lecture hall. In 1974, Jocelyn was the first person to discover spinning dead stars called radio pulsars, but she was left out when two men she worked with won a Nobel Prize for her discovery.

For centuries, women in STEM struggled to gain recognition for their work. In the 1700s, Laura Bassi was not allowed to lecture in public without the permission of men. In 1925, Cecilia Payne-Gaposchkin wrote a paper based on her research showing that the sun is made of hydrogen and helium. Two male astronomy professors told her she was wrong and then took credit for the discovery when they realized she was correct.

In 1957, the physicist Chien-Shiung Wu was ignored when a Nobel prize was awarded to two men whose theory she had confirmed by designing and performing a complicated physics experiment.

Despite the obstacles, women like these overcame significant hardships to make groundbreaking contributions to STEM fields. Their curiosity, passion, and perseverance made a difference. By reading about the modern scientists in this book, you'll see that, although things still aren't perfect, people who love science are no longer limited by their gender. Each biography also includes an activity, if you'd like to get a better feel for the science behind the stories.

Next time you picture a scientist, I hope you'll think of Raychelle Burks, Danielle Lee, Robin Wall Kimmerer, Lesley de Souza, Chanda Prescod-Weinstein, or one of the other brilliant scientists whose stories appear in these pages. Because once we move past the image of an old man with crazy hair, the sky is the limit.

Maria Sibylla Merian

BIOLOGICAL ILLUSTRATION/METAMORPHOSIS

AN ARTIST

Maria Sibylla Merian was born in Germany in 1647. She was encouraged to paint and draw by her stepfather, an artist, who specialized in painting flowers. At that time, women didn't pursue the study of nature, but painting, drawing, and embroidery were considered acceptable hobbies for girls. Maria proved exceptionally talented at drawing the plants and insects she collected.

A NATURALIST

A curious child, Maria started investigating insects when she was young. Her life took a new turn when, at thirteen years old, she learned to raise silkworms. As she observed their metamorphosis from caterpillar to moth, she wrote, "I realized that other caterpillars produced beautiful butterflies or moths, and that silkworms did the same. This led me to collect all the caterpillars I could find in order to see how they changed."

METAMORPHOSIS

Maria married and had children, but she continued to paint. Besides teaching other young women to draw, she learned to make copperplate engravings of her illustrations so that they could be printed on paper. It was important to Maria that she accurately represent the color of the plants, insects, and spiders in her work, so she hand-colored many of the prints she made herself.

CATERPILLARS

After years of work, Maria published two volumes of stunning illustrations of the life cycles of the caterpillars she'd studied. Other artists had drawn butterflies before, but she was the first to accurately depict their entire life cycles, including where butterflies and moths laid their eggs, which plants certain caterpillars ate, where they could be found, and every step of their development, including molting. Her drawings highlighted the differences between males and females of different species as well, depicting butterflies from different angles to aid in identification. The book was popular with the public but ignored by other naturalists, because she had used common names rather than Latin descriptions.

SURINAME

When she was fifty-two, Maria became the first European woman to travel independently to South America to study insects. In the country of Suriname, she observed and drew insects, reptiles, amphibians, and plants. When she returned home after three years, she published a book of illustrations based on her observations there. Many illustrators and naturalists were hugely influenced by her work and Carl Linnaeus used her drawings to identify more than a hundred new species, including a bird-eating spider, which was later named *Avicularia merianae* in her honor. Maria has several butterfly species named after her as well.

IN TODAY'S WORLD

Entomology is the modern name for the study of insects and how they relate to the environment and other living things. Insects can be both helpful and harmful. Understanding the life cycles of insects is essential to modern scientists studying everything from agriculture and medicine to solving crimes.

TRY THIS AT HOME!

Search for caterpillar eggs, caterpillars, and chrysalises. Photograph them on their host plants and draw or paint them.

• BIOLOGY •

IV.
III.
II.
I.

$C_{14}H_{24}O_9$

Fanny Hesse

AGAR GROWTH MEDIAN

'LINA

Fanny Angelina Eilshemius's friends called her "'Lina." The oldest of ten children, she was born in New York in the summer of 1850 to wealthy Dutch immigrants. There were no vaccines to prevent childhood diseases back then, and five of her siblings died when they were very young. Growing up, she learned to cook from her mother and their servants. When she was fifteen years old, Fanny was sent off to boarding school in Switzerland to study French and economics.

A PARTNER

Fanny met her husband Walther Hesse when he was visiting New York during a stint as a surgeon for a German passenger ship. They met again when her family was in Europe and became engaged. She moved to Germany in 1874 to marry him and was first exposed to science in 1884 when Walther went to work in the lab of the famous microbiologist Robert Koch.

AN ARTIST

From a family of artists, Fanny proved herself to be talented at documenting Walther's research by making colorful medical illustrations of the bacteria he studied. To create accurate drawings of individual bacteria they saw under the microscope, and clumps of bacteria called colonies that could be seen with the naked eye, she had to learn about microbiology.

A LAB ASSISTANT

Besides educating and caring for their three sons, Fanny was Walther's lab assistant. She prepared food called culture medium for the bacteria they were studying. At that time, Robert Koch, Walther, and other microbiologists were having difficulty growing individual colonies of bacteria, which severely limited their research. They mixed nutrients with gelatin to make a flat, bouncy surface where colonies grew well, but on hot days the mixture melted. Certain types of bacteria also made chemicals called enzymes, which could liquify the gelatin.

AGAR

Fanny Hesse had learned when she was young that a substance called agar-agar, or agar, made from red algae, could be added to jelly and pudding to make them resistant to melting. She had the idea to add agar to microbial culture medium and found that it created a long-lasting, heat- and enzyme-resistant surface for growing bacteria. Following her discovery, Robert Koch used Frannie's new agar culture medium recipe in his lab to grow bacteria and went on to make many great discoveries in microbiology. Sadly, Koch never gave Fanny or Walther Hesse credit for their contributions and they never profited from her brilliant idea.

IN TODAY'S WORLD

Agar growth medium is still used in labs around the world every day, in everything from molecular biology research to diagnosing strep throat in clinical labs.

* BIOLOGY *

TRY THIS AT HOME!

Make two identical desserts, one with gelatin and one with agar. Test a spoonful of each to see which one melts more easily in hot water and then eat the rest.

Susan La Flesche Picotte

PUBLIC HEALTH/HOUSEFLIES

DAUGHTER OF A CHIEF

Susan La Flesche Picotte was born in June of 1865. She was the daughter of Joseph La Flesche, who had been made chief of the Omaha tribe of Nebraska in 1853. Following years of violence and displacement by European settlers, their people divided into those who followed the traditional ways and those, like her father, who believed that the only way to survive was to incorporate the ways of the white man.

She grew up in a log cabin with her three sisters and was sent to a boarding school on the reservation where indigenous students were forced to learn about European and American culture. By the time she went off to college, Susan was fluent in English and French as well as Omaha, which she spoke with her parents.

A LIFE-CHANGING EXPERIENCE

When Susan was eight years old, she watched an elderly woman on the reservation die in pain while waiting for a white doctor. They called the doctor four times, but he never arrived. Later, she recalled thinking, "It was only an Indian, and it did not matter (to him)." Perhaps this motivated her to eventually attend medical school at the Women's Medical College of Pennsylvania, the first medical school in the United States dedicated to educating women to be physicians. Susan took biology, chemistry, anatomy, and physiology classes, and she graduated as the valedictorian of her class.

HOMECOMING

Dr. Picotte returned home to Nebraska to find patients lining up at her door suffering from diseases that included tuberculosis and cholera. She worked twenty-hour days, traveling on miles by foot, horseback, or carriage to help people. After getting married and having two sons, she continued her work, which was unusual for a woman at that time. To control diseases on the reservation, she encouraged good hygiene (cleanliness to improve human health), including fresh air and screen doors to keep disease-carrying flies out of buildings.

"DR. SUSAN"

During her medical career, Susan La Flesche Picotte served more than 1,300 patients spread out over 450 square miles. She became known as "Dr. Susan" and became a local leader. Following the death of her husband, she worked as a defender of the ancestral land belonging to the Omaha tribe and raised enough money to build a hospital on the reservation. She died in 1915.

IN TODAY'S WORLD

Public health experts today understand much more about how disease is spread than they did in Susan La Flesche Picotte's time, but they are still working to find better ways to slow or stop the spread of diseases carried by flies and mosquitoes.

TRY THIS AT HOME!

Make a model of a housefly from Play-Doh with cotton swab legs. Touch the fly's feet to an ink stamp pad and then onto paper to see how flies pick up germs and spread them around.

Mary Agnes Chase

AGROSTOLOGY/GRASSES

DAUGHTER OF A BLACKSMITH

In 1869, Mary Agnes Meara was born in Iroquois County in Illinois. Mary's father was a railroad blacksmith and her family changed their last name to Merrill to avoid the prejudice faced by working-class Irish immigrants. He died when she was only two years old and her family moved to Chicago. Mary's family was poor, so when she finished grade school, she had to go to work at the newspaper to support them and wasn't able to continue her formal education.

THE EXPOSITION

When she was only nineteen, Mary married a man in the newspaper business and changed her name to Mary Agnes Chase, but her husband died of tuberculosis a year later. In 1893, Mary and her nephew attended Chicago's Columbian Exposition, which was a huge fair, and saw an exhibit about plants. Mary was so fascinated by what she learned there that she decided to study botany.

A TALENTED ARTIST

Mary started her own field journals about plants and one of her professors was so impressed with her drawings that she was hired to illustrate books. She learned to use a microscope and worked as a meat inspector at the Chicago stockyards before returning to work with plants at the Chicago Field Museum. Eventually, she got a job with the United States Department of Agriculture, where she worked her way up from illustrator to lab assistant to principal scientist in charge of agrostology, the study of grasses. She traveled the world collecting more than 20,000 species of grasses and was the first to document several of them.

A HUMAN RIGHTS ADVOCATE

Throughout her career, Mary Agnes Chase worked tirelessly to make science accessible and understandable to everyone. Sometimes, she risked her career to speak up for human rights. In addition to mentoring underprivileged students, including women who wanted to be botanists, she was a suffragist who was arrested twice, protesting for women's right to vote.

A COLLEGE DEGREE

In her long career, Mary Agnes Chase published more than seventy articles and books, including a popular guide to grasses for nonprofessionals, called *Agnes Chases's First Book of Grasses*, which is still in print. When she was eighty-nine years old, she finally received her first college degree, an honorary degree from the University of Illinois.

IN TODAY'S WORLD

The study of agrostology is still important today. Crop plants such as rice, corn, wheat, and sugarcane are all grasses. Healthy grasslands prevent soil erosion and help with the problem of climate change by soaking up carbon dioxide from the air.

TRY THIS AT HOME!

Fold a paper towel, get it wet, and slide it into a small plastic sandwich bag so it is flat. Sprinkle grass seeds across the surface of the paper towel. Tape the bag in a sunny window and keep the paper towel damp. Watch the grass seeds germinate (sprout).

Ynés Mexía

PLANT COLLECTION/IDENTIFICATION

NATURE-LOVER

Ynés Enriquetta Juliette Mexía was born in 1870 in Washington D.C. Her father was a Mexican diplomat, and when her parents divorced, she moved to Texas with her mother. No matter where she was, whether it was Texas or Mexico, she loved taking long walks and studying the birds and plants she encountered. As a young woman, Ynés couldn't attend college, but even as an adult, she remained interested in the natural world.

CALIFORNIA

In 1909, Ynés moved to San Francisco, California. After experiencing personal tragedies, she was mentally and physically exhausted, but she found comfort in the beautiful mountains and redwood trees of Northern California. She became an active member of the Sierra Club, and signed up for a botany course at the University of California–Berkeley.

SLEEPING UNDER THE STARS

Ynés was fifty-one when she started taking classes at Berkeley, and she went on her first major plant-collecting expeditions to in 1925, when she was fifty-five. Ignoring the male explorers who said that it would be impossible for a woman to travel alone in South America, she continued taking trips up and down the Americas, wearing pants, riding horses, and sleeping under the stars. She survived an earthquake, poisonous berries, and extremely difficult conditions on her quest to discover new species and document especially interesting plants, such as the wax palm tree, which could grow up to 200 feet (61 m) tall. Ynés Mexía joined other expeditions as well, including one to Brazil headed by Agnes Chase and her colleague A. S. Hitchcock. Her assistant, Nina Floy Bracelin, prepared and helped identify the plants Ynés collected.

A SCIENCE COMMUNICATOR

Although Ynés Mexía was so busy collecting plants that she never finished her college degree, she gathered almost 150,000 species. In thirteen short years, Ynés discovered 500 new plant species and 2 new genera. Between expeditions, she gave lectures in San Francisco, shared photographs of her journeys, and taught people about plants. Ynés suffered from prejudice because of her Mexican heritage, her age, and her gender, but everyone who knew her said that she was friendly, unassuming, and tough. She died in 1938 in Berkeley, California.

IN TODAY'S WORLD

Today, specimens collected by Ynés Mexía can be found in museums from New York to San Francisco.

TRY THIS AT HOME!

Collect plants and flowers and try to identify them using a book or an app. Place the plants between pieces of newspaper sandwiched between thicker pieces of paper and press them with a heavy book or another flat, heavy object until they are dry.

Rosalind Franklin

DNA STRUCTURE

HIGH SCORER
Born in London, England, in 1920, Rosalind Franklin was the second of five children. A bright child who loved math and solved equations for fun, Rosalind was athletic as well. She played cricket and field hockey in school, although her music instructor, the composer Gustav Holst, wrote to her parents that she was not skilled in that particular area of study.

ROCKS FULL OF HOLES
In college, Rosalind studied chemistry, French, and German. After graduating, she studied the porous properties of coal, which was important in the development of charcoal filters for gas masks used during World War II. She received her PhD from Cambridge for this work.

X-RAY VISION
When the war was over, Dr. Franklin went work in a French lab that studied different compounds using X-ray diffraction, a technique that bounces energetic rays off of objects to make images. She focused her X-rays on coal and then on graphite. Soon she was an expert in a technique called X-ray crystallography, which aims X-rays at crystals to form an image on film, making it possible to determine the structure of molecules.

A TWISTED LADDER
In 1950 Dr. Franklin was working as the X-ray diffraction expert at a lab at Kings College in England, where she was attempting to take a picture of DNA to help scientists understand its structure. DNA is also called the blueprint of life because it is like a map for living things to reproduce themselves. Franklin's background in chemistry came in handy for expertly preparing DNA samples, and she captured what is now called "Photo 51": the first clear image of DNA. The clear "X" shape of the image, paired with her measurements, revealed that DNA has a spiral geometry and hinted at the distance between atoms. Soon after being shown Dr. Franklin's photograph, without her permission, scientists James Watson and Francis Crick were able to put together their famous double-helix model of DNA.

GONE BUT NOT FORGOTTEN
In her lifetime, Rosalind Franklin was given little, if any credit for her essential contribution to solving DNA's structure. Watson's descriptions of Dr. Franklin in his memoir *The Double Helix* were sexist and dismissive. She died young of cancer, and just four years after her death, Watson, Crick, and Wilkins were awarded the Nobel Prize for their work on DNA. Fortunately, today most scientists understand the essential role Franklin played in solving DNA's structure. She also made important contributions to the study of another chemical called RNA and helped decipher the structure of viruses.

IN TODAY'S WORLD
Dr. Franklin's role in the discovery of the structure of deoxyribonucleic acid (DNA) helped lay the foundation for modern molecular biology, which is our understanding of life at the molecular level.

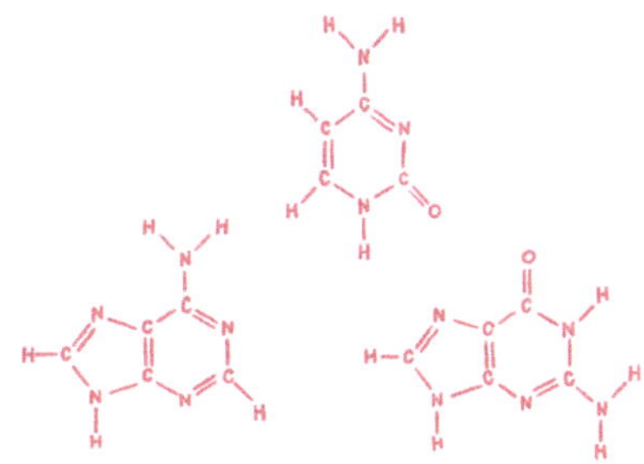

TRY THIS AT HOME!
Use pipe cleaners (chenille sticks) and cotton swabs to create a long, flexible ladder. Look at a picture of DNA structure and twist the ladder you created into a double-helical shape. Observe the double helix from one end to see the shape Rosalind Franklin photographed.

Margaret S. Collins

ZOOLOGY/TERMITES

AN EXPLORER

As a girl, Margaret Collins loved to explore in the woods near her house in West Virginia. Born in 1922, she was the fourth of five children. Her parents both attended college, and her father had gone on to receive a master's degree. Her mother, who had wanted to be an archaeologist, wasn't able to complete her undergraduate degree, but she taught her children the importance of education, a lesson that Margaret never forgot.

CHILD PRODIGY

By the age of six, Margaret's advanced intellect was already evident. She was called a child prodigy, given a library card at the West Virginia State College Library, and allowed to skip two grades. When she was only fourteen years old, Margaret graduated from high school and went to college. Graduating with a degree in biology, she also earned minors in physics and German.

STUCK IN THE LAB

Margaret completed her PhD at the University of Chicago in the lab of Alfred E. Emerson, a termite expert. Not only was Emerson an expert, but he had the largest collection of termites in the world. Emerson was a good mentor and shielded Margaret from some of the racism she faced. Unfortunately, like many male scientists at that time, he didn't want a woman doing field work with him, so while he chased down interesting termites, Margaret was stuck in the laboratory. Despite this, she managed to write an important paper on termites.

A PROFESSOR

Zoology is the branch of biology that studies the animal kingdom. When she got her PhD, Dr. Collins became the first African American woman entomologist and the third female African American zoologist. She went to work as a professor at Howard University but left because men and women weren't treated equally. She later worked for Florida A&M and spent a year at the University of Minnesota before returning to Howard University. During these years, she was fighting hard for civil rights.

GETTING INTO THE FIELD

By this time, Dr. Margaret Collins had come to think of herself as a field biologist and loved traveling the world to study termites in their natural habitat. In 1989, she discovered a new species of termite in Florida, called the Florida damp wood termite. She died in the Cayman Islands in 1996 at the age of seventy-three, doing what she loved best: studying termites.

IN TODAY'S WORLD

Scientists today still study termites, but not just because they can be destructive. Companies are attempting to duplicate termite saliva for industrial use, because it makes termite mounds so strong. Researchers are also interested in the structure of termite mounds, which stay cool inside when it's hot outdoors.

TRY THIS AT HOME!

Mix 1/4 cup (85 g) water, 1/2 cup (170 g) washable glue, and 1/4 cup (85 g) cornstarch together to make a solution resembling termite saliva. On a foil-covered baking sheet, add 2 tablespoons (28 g) termite saliva to 1/2 cup (170 g) dried coffee grounds and shape the mixture into a termite mound. Try adding different amounts of termite saliva to coffee grounds to find the best mix for creating a strong mound. Dry and compare.

Esther Lederberg

LAMBDA PHAGE/REPLICA PLATING

THE GREAT DEPRESSION

Esther Miriam Zimmer was born in the Bronx of New York in 1922. She was seven years old when the Great Depression struck. Esther recalled being hungry as a child and often having just a piece of bread, topped with juice squeezed from a tomato, for lunch. She was very close to her grandfather, who taught her to speak Hebrew.

STRUGGLING SCIENTIST

When she was sixteen, Esther graduated from high school and won a scholarship to study at Hunter College, City University of New York. She loved literature, music, and French, but decided to study biochemistry. After college, she moved to California to pursue a master's degree in genetics at Stanford. To pay rent and buy food, she worked as a laboratory assistant and did laundry for her landlady.

MADISON, WISCONSIN

In 1946, Esther married a scientist named Joshua Lederberg and took his last name. They moved to Madison, Wisconsin, where Joshua had been offered a job as a professor. Despite her master's degree and doctorate research at Stanford, Esther could only find work as Joshua's unpaid laboratory assistant. At the University of Wisconsin, she and Joshua did groundbreaking research on bacteria, and in 1959 they returned to Stanford, where Joshua was made a tenured professor. Even with a PhD and several major discoveries, Esther worked for fifteen years as a senior scientist before she was made an adjunct professor and was named the head of Stanford's Plasmid Reference Center.

A MAJOR DISCOVERY

Around 1950, as she finished her doctorate degree, Esther discovered that the *E. coli* bacteria she was studying had been infected by a type of virus called a bacteriophage. She named her discovery *lambda phage* and her research revealed that it was special because it could insert itself into the DNA of the host bacteria. While studying lambda phage, she also found a DNA sequence in bacteria that allows them to exchange DNA with other bacteria.

REPLICA PLATING

Dr. Esther Lederberg created an important laboratory technique called replica plating, which used a velvet stamp pad to transfer bacteria colonies from one plate to another in the exact same pattern. Using this method, scientists can easily find bacteria that have undergone spontaneous DNA mutations, which allow them to grow on plates containing antibiotics.

IN TODAY'S WORLD

Lambda phage is an important tool for molecular biologists and is used in research labs around the world every day.

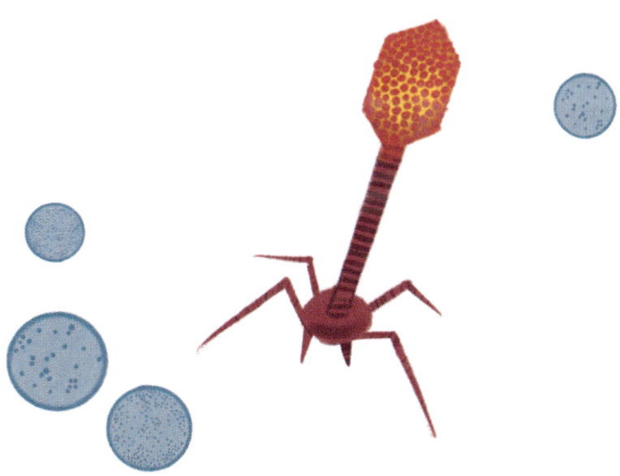

TRY THIS AT HOME!

Draw identical circles on a piece of paper. Use a stamp to transfer ink from a stamp pad to each of the circles without re-dipping in the ink. This is a model of how bacterial colonies can be transferred from one agar plate to another in the exact same pattern for research.

June Almeida

CORONAVIRUSES/AGGLUTINATION

A NEW VIRUS

When June Almeida wrote a paper about a new virus she'd discovered, scientists reviewing her paper told her that she'd made a mistake. June, who was one of the world's most skilled electron microscope technicians, would soon prove them wrong. She had, in fact, photographed and identified an entirely new kind of virus, which she and her colleagues later named coronavirus.

STRAIGHT TO WORK

Born in Scotland in 1930, June was a bright student, but she didn't have enough money to go to the university. When she was sixteen, she left school and went to work as a laboratory technician at a hospital, where she looked through a microscope for the first time. June proved extremely talented at histopathology, which involves studying slides of human tissue and identifying diseases.

ELECTRON MICROSCOPY

She eventually went to work in a lab in London, got married, and moved to Canada where she learned to use electron microscopes, which are powerful enough to magnify viruses. Although she didn't have a college degree, June quickly became so talented at microscopy that researchers she worked with made her a co-author on some important scientific papers about the structures of viruses.

CLUMPING VIRUSES

A British professor of virology named A. P. Waterson convinced June to return to England, where she was awarded a Doctor of Science degree. Now called Dr. Almeida, June became the first person to visualize the rubella virus, which causes the German measles. She also pioneered a technique called *immune electron microscopy*, which uses antibodies to clump viral particles together so that they're easier to see.

A HALO

When a colleague sent her a sample of an unidentified virus to look at with her electron microscope, June thought it looked similar to the one she'd tried to publish a paper about before. This time people believed her when she said that she'd discovered an entirely new kind of virus. She and her colleagues named this new family of viruses coronaviruses, for the crown-like halo that was visible around the viral particles in June's photographs.

Later in her career, Dr. Almeida published important papers on a bird respiratory virus and the hepatitis B virus. She also came out of retirement in the 1980s to help take images of HIV, which causes the disease AIDS.

IN TODAY'S WORLD

Thanks to the work of June Almeida and her colleagues, scientists were recently able to identify the SARS virus and the SARS-CoV2 virus that causes COVID-19 as coronaviruses. What Dr. Almeida and her colleagues discovered about the structure of these and other viruses still helps scientists develop vaccines and drugs to keep humans and animals safe from them today.

TRY THIS AT HOME!

Make models of coronaviruses by rolling Play-Doh into several small balls. Use another color to cover each ball with tiny Play-Doh spikes. These spikes form a crown-like halo in photographs and are the reason for the virus's name, since "corona" is borrowed from the Latin word for crown.

Patricia Bath

MEDICAL DEVICES/CATARACT SURGERY

A CHEMISTRY SET

Patricia Bath was born in the Harlem neighborhood of New York City in 1942. Her father was a subway train operator and her mother worked as a housekeeper when she wasn't taking care of Patricia and her brother, so she could save money for her children's education. Patricia's parents always encouraged her to work hard in school, and her mother sparked her interest in science when she bought Patricia her first chemistry set.

A SCIENCE-LOVER

Patricia was an outstanding math and science student in high school, and she discovered that she loved biology. When she was sixteen, cancer cell research she did at a workshop sponsored by the National Science Foundation was so impressive that it was included in an academic paper. After earning a BA in chemistry, she attended Howard University's College of Medicine, where she received her medical degree in 1967. Following the assassination of Martin Luther King Jr. that same year, she organized her fellow medical students to volunteer their time and talents to help people in their community who could not afford health care.

AN OPHTHALMOLOGIST

When Patricia, now Dr. Bath, returned home to work at Harlem Hospital, she noticed there were more blind patients there than they were at a neighboring hospital where she also worked. She continued her education, doing a residency to become an ophthalmologist, which is an expert in eyes and vision. When Dr. Bath's research demonstrated that certain groups of people suffered from more eye problems than others, she wanted to understand why and help address the problem. She set up an eye clinic at Harlem Hospital Center, where they started doing eye surgeries. In 1972, Patricia got married and had a daughter named Eraka.

AN INVENTOR

After moving to Los Angeles, Dr. Bath continued working to improve surgical treatment for blind patients. She invented a medical device that dissolves cloudy lenses called cataracts, which can form on the eyes of older people. Once the cataracts are gone, new lenses can be put in. She patented four more devices for eye surgeries, and thanks to her inventions, people who were blind for decades were able to regain their eyesight. Throughout her career, Dr. Bath addressed issues in society that contributed to vision problems, such as poverty and inadequate access to health care. Dr. Patricia Bath was a partner of the American Institute of Blindness, whose motto is "Eyesight is a basic human right." She died in 2019.

IN TODAY'S WORLD

Dr. Bath's contributions to surgical equipment and eye care are still used in eye clinics around the world every day.

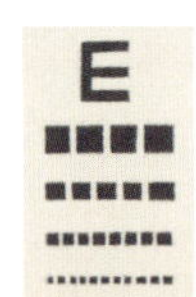

TRY THIS AT HOME!

Use waxed paper held in place by rubber bands to cover the ends of two toilet paper tubes and look through them. This will help you to understand how cloudy lenses, called cataracts, make it hard for some people to see. Replace the waxed paper with saran wrap and look through the tubes again. Now, you will be able to observe how cataract surgery improves vision by replacing the cloudy lenses with clear new ones.

E

Linda Buck

OLFACTORY CHEMISTRY

A HAPPY CHILDHOOD

Linda Buck was born in Seattle, in 1947, to parents who encouraged her creativity from the beginning. With a father who was an electrical engineer and a mother who loved puzzles, Buck was primed for a lifetime of innovative problem-solving. Given the freedom to pursue a number of interests, Linda wasn't sure what she wanted to study in college, but she was drawn to psychology, which is the study of the human mind.

A NONTRADITIONAL PATH

After attending the University of Washington, Linda wasn't sure that she wanted to be a psychotherapist, so she left school to travel but continued taking classes when she could. A course in immunology, which is the science of how our bodies fight infection, got her interested in biology, and she graduated from college with a degree in microbiology and psychology when she was twenty-eight. She moved to Texas to earn a PhD in immunology before going to work in the lab of Richard Axel at Columbia University.

YEARS IN A LAB

In Axel's neuroscience lab, Dr. Buck was fascinated by cells that send signals to the brain, and became obsessed with the idea of figuring out which cells in the nose detect scent. She also wanted to understand which genes (DNA codes) are related to these cells, and how the cells can recognize so many different odors. She worked for years to unravel the mystery, trying everything she could think of, with nothing to show for it.

MYSTERY SOLVED

Dr. Buck could hardly believe it when, in 1991, she discovered a group of genes that no one had ever seen before. These genes coded for a group of 350 smell receptors that work in combination to detect thousands of different odors. In 2004 Linda and her colleague Richard Axel shared the Nobel Prize in Physiology or Medicine for their discoveries of odorant receptors and the organization of the olfactory smell system. She said, "As a woman in science, I sincerely hope that my receiving a Nobel Prize will send a message to young women everywhere that the doors are open to them and that they should follow their dreams."

IN TODAY'S WORLD

Currently, many researchers are working to create "artificial noses" for environmental monitoring, medical diagnosis, and a host of other applications.

TRY THIS AT HOME!

Collect food items, herbs, and/or flowers with strong smells. Put the items on plates. Blindfold a friend or family member to see whether they can identify each object by smell.

Danielle Lee

PATTERN RECOGNITION/SCIENCE COMMUNICATION

LIGHTNING BUGS

Danielle N. Lee grew up in South Memphis, Tennessee. Her biological father was a musician and her stepfather, who married her mom when she was eight years old, was a tractor-trailer tire repairman. Until she married, Danielle's mother worked for Memphis Parks Commission and raised Danielle with the help of her family in a big multigenerational household. Because of her mom's job, Danielle spent lots of time outdoors, exploring woods and fields. Looking for four-leaf clovers trained her eyes to see patterns. She also loved collecting lightning bugs and making garlands from dandelions and clover grass flowers.

LOOKING FOR PATTERNS

Even as a child, Danielle was fascinated by animal behavior. When flocks of birds moved across the sky, she studied the patterns in their movements and wondered whether their flight patterns were deliberate or whether they were trying to decide which bird to follow. She was never bored. During her free time, Danielle read books, listened to music, and did crafts. She enjoyed school and was good at it, but sometimes got in trouble for talking to her friends too much. Her teachers might have been surprised to discover that Danielle's excellent communication skills would come in handy later, when she became a professor.

A PROFESSOR

Danielle earned her bachelor's degree from Tennessee Technical University in 1996. At first, she wanted to go to veterinary school, but that didn't work out. Luckily, she started to study how small mammals called voles use their sense of smell and fell in love with the discipline of mammalogy. She got a master's degree from the University of Memphis and then a PhD in biology from the University of Missouri–St. Louis. Today, Dr. Lee teaches mammalogy (the study of mammals) and urban ecology at Southern Illinois University–Edwardsville. The focus of her current research is the behavior of African giant pouched rats, and she loves traveling to Tanzania to do field studies.

A SCIENCE COMMUNICATOR

Dr. Danielle N. Lee is well known for her ability to excite people about science. She does outreach to the public on social media, but she is especially passionate about her work as a role model and educator for underserved audiences. As a woman of color in science, she has faced extraordinary challenges and persevered. Dr. Lee has received many honors for her work, especially for encouraging minority participation in Science, Technology, Engineering, and Math (STEM) fields. She is a 2015 TED Fellow and a 2107 National Geographic Emerging Explorer.

THE FOUR-LEAF CLOVER

The logo for Dr. Lee's laboratory is the four-leaf clover. Each leaf is associated with a word: Behavior, Ecology, Justice, and Outreach. The study of science depends on observation, and the symbol reminds people who work with her that a multitude of perspectives are essential. She says that the clover represents why we need different perspectives: just as different patterns stand out to different individuals looking at the exact same field of clovers, a multitude of viewpoints bring the big picture into sharper focus.

IN TODAY'S WORLD

Research on mammals and animal behavior is essential to gaining insight into everything from economics to human health and wildlife conservation. Modern scientists, such as Dr. Lee, understand that it is essential to communicate with the public to gain support for research and to inspire future scientists.

TRY THIS AT HOME!

Hone your pattern recognition skills by hunting for four-leaf clovers. Clover plants are part of the genus *Trifolium*, which means three leaves, but sometimes the DNA in clover undergoes genetic mutation and grows an extra leaflet.

Rae Wynn-Grant

CARNIVORE ECOLOGY/ANIMAL BEHAVIOR

URBAN CHILDHOOD

Rae Wynn-Grant grew up in big cities. Born in San Francisco, she spent most of her childhood in California. Her mother is a writer and her father is an architect, so it surprised everyone when she decided to be a wildlife biologist. As a kid, she remembers being fascinated by television shows about wild animals. When she was a teenager, Rae dreamed about hosting her own nature show on *National Geographic*. In middle school and high school, she struggled with traditional tests and found math and science challenging, but Rae loved the things she was learning about the natural world and didn't allow less-than-perfect grades stand in her way.

ENVIRONMENTAL STUDIES

After high school, Rae attended Emory University where she learned about conservation biology, which focuses on protecting and restoring the diversity of life on Earth. She didn't get to see her first wild animal until she traveled to East Africa when she was nineteen years old. Rae studied for a master's degree in environmental studies from Yale University and then got a doctorate degree, studying how carnivores adapt (change) their behavior to adjust to landscapes altered by humans.

LIONS AND LEMURS AND BEARS

Rae loves bears. "They're just like people," she said in an interview with National Geographic Explorers at Work. "They just love to hang out." After finishing her PhD in ecology and evolution from Columbia University, she did research called a post-doctoral for the American Museum of Natural History, studying the behavior and habitat of grizzly bears in Montana.

Today, she studies bears in the mountains of Nevada and New York and does research on lions in Tanzania and lemurs in Madagascar. Dr. Wynn-Grant is interested in learning how animals, such as lions, change their movement and hunting patterns when humans live nearby. Her science work is dedicated to helping people and carnivores coexist peacefully all around the world.

A DREAM-COME-TRUE

Rae currently works as a large carnivore ecologist for the National Geographic Society and studies carnivore conservation around the world. She is a Visiting Scientist at the American Museum of Natural History and a professor at Columbia University and Johns Hopkins University. Dr. Wynn-Grant does some of her most important work as a science communicator, making environmental science accessible to broad and diverse audiences. Rae wants to show urban kids that they have a place doing work like hers. "Don't confuse performance with passion" is her motto. Dr. Rae Wynn-Grant is living proof that passion and hard work can make dreams come true.

IN TODAY'S WORLD

As Earth's population grows, people are moving into the habitat of wild animals. The work of Dr. Rae Wynn-Grant and other conservation ecologists is essential to the well-being of humans and animals. Her dream of hosting a nature show came true in 2023, when she became the co-host of *Mutual of Omaha's Wild Kingdom*.

TRY THIS AT HOME!

Observe wild animals such as birds, squirrels, and rabbits in your neighborhood. If you have binoculars, use them to take a closer look. Start a journal to record when and where you see certain animals, what they eat, and how they interact with humans and other animals.

Eunice Newton Foote

GREENHOUSE GASES

A FAMOUS RELATIVE

Eunice Newton was born in 1819 in Goshen, Connecticut. She had six sisters and five brothers, and her father was a distant relative of the famous scientist Sir Isaac Newton. Eunice grew up in Bloomfield, New York, and attended Troy Female Seminary, which is now called the Willard School. She took chemistry and biology classes at a nearby science college, where she learned about physics and chemistry.

SUFFRAGETTE

Newton married Elisha Foote, who was an inventor, mathematician, and judge, and they had a daughter who would grow up to be an artist, writer, and social activist. In 1848, Eunice attended the first women's rights conference in America and signed a document called the "Declaration of Sentiments." Written by Eunice's friend Elizabeth Cady Stanton, the declaration called for equality with men and the right for women to vote.

AT-HOME SCIENTIST

During the nineteenth century, only sixteen physics papers were published by American women and two of those papers were written by Eunice Newton Foote. Although she didn't have a lab, Eunice did science at home. She was interested in many things, including how sunlight interacted with different gases. Using an air pump, glass jars, and thermometers, she compared how the temperature inside the jars changed when they were filled with dry air, moist air, carbon dioxide (CO_2), or hydrogen and placed in the sunlight.

HEATING AND COOLING

Her research showed that trapped carbon dioxide gas caused the temperature to rise the most. Eunice wrote, "The receiver containing this gas became itself much heated—very sensibly more so than the other—and on being removed [from sunlight] it was many times as long in cooling." She also found that the heating effect of moist air was greater than that of dry air and concluded, "An atmosphere of that gas [CO_2] would give to our Earth a high temperature."

GREENHOUSE GAS AND CLIMATE

Eunice's paper, "Circumstances Affecting the Heat of the Sun's Rays," was summarized in several journals. Unfortunately, most of them ignored her conclusion that gases we now call "greenhouse gases," such as carbon dioxide, can impact Earth's climate. Her research on sunlight, trapped gases, and water vapor was groundbreaking. Although she was the first person to propose a connection between carbon dioxide and climate change, she wasn't recognized for this accomplishment until 2010, when a geologist named Ray Sorenson came across her work in an old journal.

AN INVENTOR

Eunice Newton Foote continued to study gases and later published a paper on atmospheric pressure and electrical charges. In addition to scientific research, she invented things with her husband. Their inventions included a new type of paper-making machine and a squeakless sole for shoes, made from a single piece of rubber.

IN TODAY'S WORLD

The buildup of greenhouse gases in Earth's atmosphere from human activities such as burning fossil fuel has dramatically warmed our planet and is one of the biggest challenges we face today.

TRY THIS AT HOME!

Make a list of ways you can reduce the amount of carbon dioxide released into the atmosphere by burning fossil fuels. For example, turning off lights, carpooling, walking, and biking can all reduce your "carbon footprint."

Dora Priaulx Henry

BARNACLES

A BIG MOVE

Dora Priaulx was born in Maquoketa, Iowa, in 1904. Both of her parents were teachers. After the death of her father, when she was five, her mother moved Dora and her brothers to Los Angeles. School came easily to Dora and she graduated from Hollywood High School before getting a biology degree from the University of California at Berkeley, where she married a fellow student named Bernard Henry.

FROM PARASITES TO BARNACLES

After completing her PhD in parasitology at UC Berkeley, Dora and Bernard moved to Seattle, where she began to study parasites called gregarines, which were infecting barnacles in Puget Sound. As she worked, her interest shifted from the parasites to the barnacles themselves. She started collecting barnacles, identifying them and writing papers about the fascinating invertebrates, which attach headfirst to boats, rocks, whales, crustaceans, and almost anything else in the water.

WORLD WAR II

The Smithsonian Institution in Washington, D.C., was impressed by Dora's expansive knowledge and encouraged her to study their museum barnacle collections. She was also an important consultant for Ed Ricketts in his early work on intertidal ecology and is mentioned in *The Log from the Sea of Cortez*, which he wrote with the famous author John Steinbeck. Barnacle growth can present a big problem for ships, and World War II led Dora to work as an assistant oceanographer with the air force and then to the navy's Hydrographic Office.

TRAVEL ABROAD

Interested in the distribution patterns of barnacles, Dora was the first to accurately plot the range of places where they grew, on a map. She continued to study barnacles until she was ninety-three years old. Her fieldwork took her from Lower California and the Gulf of California to the Gulf of Mexico, the Central Pacific Islands, and Bermuda. She discovered new species, learned new things about barnacle reproduction, and wrote countless papers.

AN EDUCATOR

Dr. Dora Priaulx Henry loved to travel. In 1968, she helped develop a "Coasts of the World" travel course, which took students and faculty to the South Pacific and South America to learn about wind, waves, tides, volcanoes, glaciers, beaches, and life in the intertidal and near-shore environments. She always welcomed visits from colleagues from around the world. Today, her barnacle collection can be found at the Smithsonian Institution's Museum of Natural History.

IN TODAY'S WORLD

Barnacles still present a challenge to boats, but researchers are very interested in the strong glue they make to cement themselves to wet objects very quickly. One group of researchers has copied their oily adhesive to make a substance that can be applied to wounds to quickly stop bleeding.

TRY THIS AT HOME!

Tie several short pieces of yarn to the end of a stick to represent barnacle legs. On the bottom of a paper cup, draw two long curved lines on opposite sides. Connect them in the center to create what looks like the letter "I" with curved serifs. Cut along the lines to make a flapped door, representing the calcium plates on barnacles. Push your barnacle legs up and down through the door to show how barnacles emerge to feed and retreat from predators.

* ECOLOGY *

Rachel Carson

ENVIRONMENTAL CONTAMINANTS

A NATURE LOVER

Rachel Carson always wanted to be a writer and published her first article for a children's magazine when she was eleven. Born in Springdale, Pennsylvania, in 1907, she loved reading and exploring nature. The fields and woods near her house were filled with birds and other animals, although on some days, everything was coated with gray ash from steel mills downriver from where she lived. Her family struggled to pay the bills, and Rachel's mother gave piano lessons to help support Rachel and her two siblings.

AWAKENED BY BIRDS

Every morning, Rachel woke to hear the birds singing. She worked very hard in school, and in her senior paper for high school she wrote that a "thinking, reasoning mind" was a person's most valuable possession. Her parents sold off part of their land to help pay for her college. Rachel got a master's degree from Johns Hopkins University and spent time at the Marine Biological Laboratory in Woods Hole, Massachusetts, where she explored tide pools and fell in love with the sea.

AN AUTHOR

After graduating, Rachel found a job as a writer for the U.S. Department of Fish and Wildlife. Finding inspiration in the natural world, she started to write books about the sea, weaving in geology, chemistry, and biology. With words, she painted beautiful images of the intertidal zones and the open ocean, making science engaging and easy to understand. Rachael organized her descriptions of marine ecosystems by zones that started on shore and moved to deeper water. By 1962, she was a best-selling author.

DDT

Near where Rachel worked in Maryland, a new pesticide (insect killer) called DDT was being tested for safety. Her job at the Department of Fish and Wildlife made her a witness to some of the first reports that DDT was also deadly to fish and wildlife populations. Her background in science helped her understand that DDT could move through soil and water from the fields and wetlands where it had been sprayed into nearby lakes, streams, and rivers. The chemical moved up the food chain, from invertebrates into the bodies of fish, birds, and mammals, often killing them or making it harder for them to reproduce.

AN ACTIVIST FOR NATURE

In 1962, Rachel sparked the environmental movement. She noticed the gradual disappearance of birds after the government started spraying DDT around the United States, and she wrote a book called *Silent Spring*. The book was a best-seller and used scientific evidence to show that DDT was harming ecosystems. The huge corporations that produced DDT called Rachel "hysterical," and "an uninformed woman speaking of that which she did not know." But she refused to be silent and in 1972, the use of DDT was banned in the United States.

IN TODAY'S WORLD

Rachel Carson reminded us that what we do to the environment, we do to ourselves. Many of today's ecologists study the impacts of human behavior on Earth's ecosystems, searching for ways we can maintain healthy relationships with other living things.

TRY THIS AT HOME!

Dip toothpicks in food coloring, poke them into clear or yellow Jell-O or gelatin and refrigerate for 24 hours. Over time, the dots of food coloring will spread out as the dye moves through the gelatin. This illustrates how chemicals such as pesticides disperse through soil.

Sylvia Earle

OCEAN RESEARCH

POND EXPLORER

The ocean first got American oceanographer Sylvia Earle's attention when she was three years old, and a wave knocked her off her feet. She's been running back into the water ever since. Born in 1935, Sylvia lived on a farm in New Jersey with her parents and two brothers until she was thirteen. A nature lover, she spent her free time exploring the life in their backyard pond, catching fish and tadpoles to keep in jars where she could observe them more closely. When her family moved to Florida, her interest immediately shifted to the wildlife in the Gulf of Mexico.

SCUBA

Sylvia was such a dedicated student that she graduated from high school when she was sixteen and got a scholarship to Florida State University. Interested in marine plants, she became a certified SCUBA diver to study her subjects close-up. Sylvia went to graduate school at Duke University and focused her research on plants called algae, which live in the water and turn carbon dioxide and sunlight into oxygen and sugars. After pausing her education to have two children, she completed her PhD in botany (the study of plants), collecting more than 20,000 samples of algae from the Gulf of Mexico. She was one of the first biologists to use SCUBA to document ocean organisms.

LIFE UNDER WATER

It was unusual for women to join ocean research expeditions at that time, but Sylvia joined several, accompanying research crews on voyages to the Galápagos Islands, the Chilean coast, and the Panama Canal Zone. She did all this while raising two children, taking classes, and writing her PhD dissertation. Sylvia continued to dive on expeditions around the world and was the first woman to set foot in an underwater habitat where scientists lived 100 feet (30.5 m) below the surface. In 1970, she led an all-female team to live and dive 50 feet (15 m) under water in the Tektite II Project, near the U.S. Virgin Islands.

A DEEP DIVE

Sylvia dove with sperm whales for the 1980 film *Gentle Giants of the Pacific*. In 1979, she famously rode a submarine to the ocean floor off the coast of Hawaii. Wearing a pressurized "JIM" suit, she was carried down 1,250 feet (381 m), deeper than anyone has ever walked the ocean floor untethered. Detaching from the submarine, she walked along the bottom for two and a half hours. Far from the sun's light, the water was inky black and she was surrounded by glowing bioluminescent creatures. Sylvia said that the experience was like diving into a galaxy of lights.

NOAA

Sylvia was the first woman to be chief scientist of the U.S. National Ocean and Atmospheric Administration (NOAA). In 1998, she was named *Time* magazine's first Hero for the Planet. Passionate about protecting the ocean, Sylvia wants people to understand how important it is to all of us. She says, "You should treat the ocean as if your life depended on it, because it does!" and reminds people that "with every breath they take, with every drop of water they drink, the ocean is touching them."

IN TODAY'S WORLD

Dr. Sylvia Earle continues to educate the public about our oceans and leads Mission Blue, which is working to inspire "an upwelling of public awareness, access and support for a worldwide network of marine protected areas."

TRY THIS AT HOME!

Fill a bowl with water and add a few drops of blue food coloring. Put one to two tablespoons (14–28 g) of vegetable oil on the surface of the water to represent an oil spill. Float a feather on the surface of the oil to see what happens to birds when oil spills occur. Experiment with different techniques for removing oil from the water.

Wangari Maathai

GREEN BELT MOVEMENT

KIKUYU KNOWLEDGE

Wangari Maathai was born in 1940 and grew up in the Kikuyu community in the central highlands of Kenya, surrounded by natural beauty in the countryside. She recalled her mother telling her not to collect firewood from fig trees, which were sacred to their people. When she got older, Wangari realized that those special trees protected the sloping landscape.

A DOCTORATE DEGREE

After attending primary school in Kenya, Wangari went to Mount St. Scholastic College in Kansas, where she majored in biology. Next, she earned a master's degree in biology from the University of Pittsburgh, in Pennsylvania. Returning to Africa, she attended the University of Nairobi and became the first woman in East Africa to earn a doctorate degree.

A CHANGED LANDSCAPE

While in the United States, Wangari was greatly influenced by the civil rights movement, and in graduate school she became a member of the National Council of Women. At one of their meetings, she learned what was happening to the environment in her beloved Kenya, which had changed dramatically over the five years when she was in America. After gaining independence from British rule, Kenya's newly independent government embraced commercial agriculture, cutting down trees to plant cash crops and build new housing developments.

DESERTIFICATION

Trees had been cut down, streams had been drying up, the water was polluted, and harvests had not been good. Women had to walk long distances to find firewood for cooking and boiling water and worried about feeding their families. Wangari understood that trees hold soil and water in place after it rains. She recognized that many of the problems in her country were the direct result of deforestation. Agriculture and development were turning fertile land into desert.

THE GREEN BELT MOVEMENT

To help solve the environmental problems, Wangari helped found the Green Belt Movement, which promised to plant millions of trees throughout Kenya. She understood that planting trees would help restore ecosystems and provide food and fuel for communities. She also fought against cutting down more forest for building and housing developments. Much of the deforestation had been done illegally, with the blessing of a corrupt government. Wangari was threatened, beaten, and even sent to jail, but she never gave up and mobilized thousands of women and men to plant trees.

52 MILLION NEW TREES

Through the simple act of planting trees, Wangari Maathai understood that ecology is more than a scientific discipline. It is a matter of war, peace, human rights, and the survival of species, including ours. In 2004, she was awarded the Nobel Peace Prize. The Green Belt Movement has already planted more than 52 million trees. Wangari died in 2011.

IN TODAY'S WORLD

Trees are very efficient at removing carbon dioxide from the atmosphere. With climate change, it is now more important than ever that we stop deforestation and plant trees to replace those that have been cut down.

TRY THIS AT HOME!

Look up information about tree seeds found in your area that are easy to grow, such as maple seeds. Find a tree seed and plant it in a flowerpot. Keep the soil damp so it can germinate and grow. When it is large enough, plant the seedling outdoors in a spot where there is plenty of room for it to grow.

Robin Wall Kimmerer

PLANT ECOLOGY/TRADITIONAL KNOWLEDGE

THE HEART BERRY

Born in 1953, Robin Wall Kimmerer grew up in Upstate New York. In English, her Potawatomi name means "Light Shining through Sky Woman." The Potawatomi Nation originally lived in the Great Lakes region of the United States, but in the late 1700s, many Potawatomi were forcibly removed by the government to a reservation in Oklahoma. It was nature that helped Robin and her parents reconnect with their Potawatomi heritage. As a girl, she loved exploring the fields and woods and wrote that "it was the wild strawberries, beneath dewy leaves on an almost-summer morning, who gave me my sense of the world, my place in it." According to Robin, "In Potawatomi, the strawberry is 'ode min,' the heart berry. We recognize them as the leaders of the berries, the first to bear fruit."

MOSS

Robin earned a master's degree in botany and then a PhD in plant ecology from the University of Wisconsin, Madison. While studying forest ecology in graduate school, she became fascinated by mosses, which are small plants with no roots. Studying them close-up, she discovered an entire miniature world. After teaching in Kentucky, Robin moved back to Upstate New York to become a professor at State University of New York College of Environmental Science and Forestry (ESF), where she had received her bachelor's degree.

WAYS OF KNOWING

Today, Robin is interested in educating her students and the public about science and traditional ecological knowledge. She wrote an award-winning book called *Gathering Moss: A Natural and Culture History of Mosses*, which beautifully weaves plant ecology into Robin's understanding of traditional knowledge, or "ways of knowing." Robin is also the director of the Center for Native Peoples and the Environment at ESF. The program gives Indigenous students more opportunities to study environmental science and allows traditional science to benefit from the wisdom of Native philosophy. Robin hopes that the center will help us reach the common goal of sustainability.

WORDS MATTER

Dr. Robin Wall Kimmerer thinks that language is extremely important. As a writer and a scientist, she is interested in "not only the restoration of ecological communities, but restoration of our relationships to land." She believes that our words affect our behavior and our attitude about the world. Rather than referring to the Earth as a thing, or "it," she suggests that we refer to our planet and everything natural thing on it as family, or "kin." She hopes that once we stop seeing nature as separate from ourselves, we will take better care of our world.

A MOTHER, A SCIENTIST, A WRITER, AND A MEMBER OF A FIRST NATION

Robin writes, "Knowing that you love the Earth changes you, activates you to protect and defend and celebrate. But when you feel that the Earth loves you in return, that feeling transforms the relationship from a one-way street into a sacred bond." Her identity as a mother, a scientist, a writer, and a member of a First Nation has made her a social activist for the environment, sustainability, social justice, and Native American issues.

IN TODAY'S WORLD

Dr. Robin Wall Kimmerer's award-winning book, *Braiding Sweetgrass*, beautifully illustrates the interdependence of humans and the natural world.

TRY THIS AT HOME!

Take a magnifying glass along and go on a hunt for velvety green and brown mosses. Mosses can be found almost everywhere: in yards, in woods, on old shingles, and even on rocks. Observe the mosses you find under the magnifying glass to get a closer look at their structure.

Dana Bergstrom

ANTARCTIC RESEARCH

NATURE LOVER

In Sydney, Australia, Dana Bergstrom knew she wanted to be scientist when she was five years old. After school, she'd climb trees and create miniature African landscapes populated with plastic lions and elephants. Sometimes, Dana's mom took her and her sister to Manly Beach to explore tide pools. Dana also loved pouring over the photographs in *National Geographic* and an atlas of the world, which contained drawings of plants, minerals, and dinosaurs.

A WOMAN IN THE FIELD

By the time she was twelve, Dana knew that she wanted to be an ecologist. After high school, she pursued her dream by completing undergraduate and graduate degrees at Macquarie University. At that time, there were mostly men in the Australian Antarctic program, but she forged a new path as one of the first female scientists to travel south to do fieldwork over a long period of time.

AN ENDANGERED ECOSYSTEM

To do research for her master's degree and PhD, Dana traveled to Macquarie Island in the southwestern Pacific Ocean. About halfway between New Zealand and Antarctica, it is one of the cloudiest places on Earth. Part of the Antipodes Subantarctic Islands tundra ecoregion, Macquarie Island is home to the world's entire royal penguin population during their annual nesting season and has an interesting ecological history. Dana was focused on studying unique vegetation on the island, but while she was there, other native species were struggling for survival due to invasive rats, cats, and rabbits brought to the island by European ships.

NEVER WAKE A SLEEPING FUR SEAL

One day, while doing research on a remote, volcanic subarctic island named Heard Island, Dana had a close encounter with a grumpy male fur seal, whom she accidentally woke from a nap. Dana stood her ground and shouted as the enormous animal charged, but the seal wasn't deterred and kept coming toward her, snorting and foaming at the mouth. Deciding to run, Dana slipped in an elephant seal wallow and fell into a smelly, slimy depression in the ground. Luckily, this brought her to a lower level than the fur seal. After looking at Dana with a satisfied face, the fur seal rocked back and forth from flipper to flipper as if he were laughing and then returned to his sleeping spot and resumed his nap.

THE 3AS

Dr. Dana Bergstrom works tirelessly to educate the public about the importance of caring for Antarctica. She has identified what she calls the 3As: awareness, anticipation, and action, which are all crucial to conserving the natural world. Dana also used her artistic skills to write a musical called *Antarctica, Beneath the Storm*, which is about a female penguin biologist who travels to Antarctica and experiences the realities of climate change. In 2021, she won the 2021 Eureka Prize for Leadership in Innovation and Science.

IN TODAY'S WORLD

Today, as a senior researcher at the Australian Antarctic Division, Dana is interested in identifying any risks against Antarctic and sub-Antarctic ecosystems and finding ways to keep ecosystems from collapsing in the future. She helped design a biosecurity center at a wharf that serves as Australia's gateway to Antarctica to help keep invasive plant and animal species out.

TRY THIS AT HOME!

Sprinkle some grass seeds on a concrete sidewalk or driveway. Get the bottom of your shoes muddy and walk over the grass seeds to see how many seeds stick to your shoes. This exercise demonstrates how easily humans can spread plants from one place to another.

Aparajita Datta

SEED DISPERSAL

A PASSION FOR BOOKS AND ANIMALS

Aparajita Datta was born in Kolkata, the capital of the Indian state of West Bengal, in 1970. When she was eight, her family moved to Lusaka, Zambia, in southern Africa. Her interest in animals was first inspired by the books of James Herriot and Gerald Durrell and one of her teachers invited her to join the Zoo Club at school. After five years, Aparajita's family returned to India, where she finished high school and studied botany at Presidency University.

THE PAKKE TIGER RESERVE

After graduating from college, Aparajita earned a master's degree from the Wildlife Institute of India. While she was in graduate school, she met Charudutt Mishra, who was also studying wildlife ecology, and they married. In 1995, she went to the Pakke Tiger Reserve in Northeast India to study how logging was affecting wildlife. While she was there, Aparajita became fascinated by hornbill birds and continued to study them for her PhD.

HORNBILLS

Hornbills are beautiful birds with long bills that sometimes look like the horn of a cow. Their beaks are often very colorful and sometimes have a bony protrusion on top. Aparajita's research revealed that hornbills are more than just pretty birds. The birds she studied in the Pakhui Wildlife Sanctuary turned out to be vital participants in the ecosystems they inhabited, spreading the seeds of more than eighty species of trees. In fact, some tree seeds were only spread by hornbills. Aparajita calls hornbills the "farmers of the forest."

SEED DISPERSAL

"Understanding connections and interactions between animals and plants is the part of my work I find most exciting," Aparajita says. "In tropical forests, 80 to 90 percent of tree species bear fruits that animals disperse. But . . . many large mammals and birds like the hornbill—crucial to seed dispersal—are hunted. Some parts of the park have become empty forests, devoid of wildlife. The absence of these dispersers could have severe consequences for the regeneration of many plant species."

PROTECTING HORNBILLS

Aparajita continued to study wildlife in the Arunachal region, which includes the Pakke Tiger Reserve. She did a wildlife census in the area, counting tigers, bears, clouded leopards, and deer in the park. She studied hornbills as well, because they were in grave danger of extinction due to hunting and the cutting down of trees where they nested. Aparajita started to work with the Lisu and Nyishi people in the area, enlisting former hunters to help protect the birds and their nests.

LEARNING FROM THE LISU PEOPLE

Passionate about nurturing people's relationship with the wild, Aparajita set up a nest adoption program. She formed a partnership between people in the cities who offered financial support and people who live in the forest, who protected hornbills. She said, "The Lisu people are right by our side. They've shown and told me things I never would have otherwise known. I think wildlife biologists often forget how much we depend on the insight of local people. To me, part of the wonder of this incredible place is being there with the Lisu, sharing moments in the forest with them." In 2016, the Pakke Tiger Reserve won an India Biodiversity Award for its Hornbill Nest Adoption Program.

IN TODAY'S WORLD

In addition to studying and helping to protect hornbills, Dr. Aparajita Datta is working on projects to restore native tree species to areas where they have been cut down.

TRY THIS AT HOME!

Photograph insects on plants and birds eating fruit and seeds. Think about why certain insects and birds might be associated with specific plant species.

Lisa Schulte Moore

PRAIRIE STRIPS

A NATURE LOVER
Lisa Schulte Moore grew up in Wisconsin, where she loved camping and visiting her great-grandmother's farm outside of her hometown of Eau Claire. After high school, she attended the University of Wisconsin, Eau Claire to study biology. "From there," she says, "it was kind of just following my interests and passions from one question to another." Her curiosity led her to northern Minnesota, where she did research for a master's degree, studying bird populations affected by wildfires and logging.

BUILDING BRIDGES
Eventually, she completed a PhD in forestry at UW and took a job with the U.S. Forest Service. While she was there, she learned that wildlife biologists and foresters sometimes had problems talking to each other, which makes it harder for everyone to do their jobs of conserving the forest. Realizing the importance of communication, Lisa decided to build bridges between groups who had different goals and priorities as part of her work going forward.

THE CORN BELT
In 2003, Lisa joined the faculty at Iowa State University, in Ames, Iowa, as a landscape ecologist with a goal of making agriculture work for both people and the planet. Iowa lies in the heart of America's Corn Belt, which was first inhabited by a number of tribes, including Dakota, Ho-Chunk, Ioway, Otoe, Illiniwek, Meskwaki, Omaha, and Sauk. When the U.S. government forced the Indigenous people off the land in the 1800s, most of it was settled as farmland.

RUNOFF
Unfortunately, farming can be harmful to the environment. Plowing up prairie to plant crops destroys ecosystems where pollinators, birds, and other animals live. Fertilizer and chemicals that are sprayed on crops to kill insects and stop weeds from growing run off the plants, into the soil, and make their way into the water supply. Lisa understands that while humans need the food, fiber, and fuel produced by agriculture, we must find ways to protect the environment.

PRAIRIE STRIPS
Lisa points out that where corn and soybeans are grown, the fields are bare for most of the year. Without roots holding it down, the soil is vulnerable to erosion by wind and water. Pollution from erosion moves into streams and rivers, ruining the water quality all the way from the Mississippi River to the Gulf of Mexico. Farm fields can also produce greenhouse gases. A prairie strip is an agricultural conservation technique that uses native grasses and plants planted in farm fields to help build soil and clean up rainwater.

CLEANER WATER, BETTER SOIL
Prairie strips also help keep fertilizer and other chemicals out of waterways while providing a habitat for native wildlife and pollinators. Lisa and her team have shown that by covering 10 percent of a crop field with prairie strips, they can keep 95 percent more of the soil in that field, 77 percent more phosphorus, and 70 percent more nitrogen, so less fertilizer is needed.

IN TODAY'S WORLD
Dr. Lisa Schulte Moore hopes that what scientists and farmers in Iowa are learning about using native plants to increase production and decrease environmental damage will someday be used in farms around the world.

TRY THIS AT HOME!
Trace the metal cap on a mason jar lid on brown paper. Cut a 0.2-inch (0.5-cm) hole in the center of the paper and then cut a line to the hole. Using pipe cleaners (chenille sticks) make a model of a small prairie plant with long roots. Slide it into the center of the brown paper circle so the roots go down. Put the roots into the mason jar and screw a ring lid on, so that the plant sits on top of the jar.

Lesley de Souza

CONSERVATION BIOLOGY

ICHTHYOLOGY

Born in 1977, Lesley de Souza grew up exploring forests and creeks in the southeastern United States and on her family's farm in Brazil. As a child, Lesley and her little brother spent hours on end in forests and creeks. In college, she fell in love with fieldwork, which allowed her to continue exploring streams and rivers. She went on to graduate school and got a PhD in neotropical ichthyology (the study of fish in Central and South America).

THE WORLD'S LUNGS

Lesley's fascination with fish took her to the Amazon Basin and a region called the Guiana Shield in South America, all part of the Amazon rain forest, the largest, most diverse tropical rain forest on Earth. Many people refer to the Amazon as the world's lungs because its trees and plants play an essential role in keeping our atmosphere stable and breathable. The Amazon is a biodiversity hot spot, containing thousands of keystone species whose populations keep ecosystems healthy.

ARAPAIMA

Working with Indigenous people who understand their environment better than anyone else, Lesley studies a fish called arapaima, which can grow up to 10 feet (3 m) long and are strong enough to break your bones if you don't handle them correctly. By installing tiny radio transmitters in the fish, she tracks their movements as they migrate into the flooded forest during the rainy season.

A RAIN FOREST ECOSYSTEM

Lesley says, "There's no way to separate what's happening underwater from the top of the forest canopy. The macaws perched on top of the giant trees are dependent on the fish swimming in the river and vice versa. The river brings nutrients into the soils that are absorbed through the roots and permeate the entire landscape . . . fish feed upon the fruits that fall from the trees."

CONSERVATION FOR THE FUTURE

Like many scientists, Lesley is deeply concerned about ecological destruction in the Amazon. Logging and mining are threatening the fragile ecosystems. Cutting down trees and digging for gold pollutes the air and water and destroys waterways. In addition to studying fish, Lesley is interested in working with the Indigenous people and governments of the places she studies to create conservation areas. This helps to ensure a healthy future for the water, the land, and the people who live there.

IN TODAY'S WORLD

Dr. Lesley de Souza understands that it is essential to address environmental issues on a local level by working closely with governments and people who live in threatened ecosystems. Every ecosystem plays an important role in the larger ecosystem that is Earth.

TRY THIS AT HOME!

Make a terrarium! Fill the bottom of a large, clear container with gravel, charcoal, dried moss, and soil. Add plants and water. Cover the container with a lid or plastic wrap so water will evaporate, condense, and drip back into the soil.

Ayana Elizabeth Johnson

MARINE CONSERVATION/PUBLIC POLICY

A GLASS-BOTTOM BOAT

Ayana Elizabeth Johnson was born in Brooklyn, New York, in 1980, and grew up there. Ayana had always been interested in animals and spent hours digging up worms in her backyard. When she was five, Ayana fell in love with the ocean. Her family had taken a trip to Florida and ventured out in a glass-bottom boat. Staring down through the transparent floor, she was captivated by the magical coral reef below. By the time Ayana was ten, she also wanted to be a lawyer so that she could help people who were fighting for social justice.

BETTER FISH TRAPS

After high school, she studied environmental science and public policy at Harvard University. While pursuing her PhD in marine biology at the University of California's Scripps Institution, Ayana traveled to the Caribbean islands of Curacao and Bonaire to do research. Working with local fishermen and divers, she redesigned fish traps to reduce bycatch, which are fish caught by accident.

COLLABORATION

The collaboration was a success, and the new traps reduced bycatch by around 80 percent, allowing fishermen to catch what they needed while keeping other fish in the coral reef ecosystems safe. Ayana's work on the islands convinced her more than ever that she wanted to take an active role in conservation. She then went on to work with the people of the island of Barbuda as they created their own marine regulations.

CHANGE MAKER

In 2016, Ayana moved back to Brooklyn, where she worked with Greenpeace to protect coral reefs and with the World Wildlife Fund. She started a think tank for coastal cities called the Ocean Collective. She also worked on an ocean-centric plan for reducing the amount of carbon we put into the atmosphere, and launched a podcast called *How to Save a Planet*.

ENVIRONMENTAL JUSTICE

Ayana points out that environmental protection is a justice issue, saying, "If you think about the rates of asthma in inner-city communities that are near power plants or exposed to other types of pollution, it's a lot higher. And when we think about immigration, and how a lot of migration is now driven by climate change, that becomes a social justice issue that was triggered by the impacts on communities that did the least to emit the carbon to cause the problem."

RETURNING THE FAVOR

As a science communicator, Ayana likes to point out the role played by the ocean in slowing climate change. She says, "The ocean has already absorbed about 30 percent of the excess CO_2 that we've trapped by burning fossil fuels. And the ocean has already absorbed 93 percent of the heat that we've trapped. And so the ocean is trying its best to buffer us from our worst. We need to return the favor."

IN TODAY'S WORLD

To protect the ocean, Dr. Ayana Elizabeth Johnson suggests avoiding plastic containers, because plastic waste often ends up polluting the ocean. Keep your carbon footprint small by walking, biking, and using public transportation. Carpooling also reduces the amount of carbon dioxide gas that is soaked up by the ocean.

TRY THIS AT HOME!

To illustrate how excess carbon dioxide in the atmosphere affects ocean water, boil red cabbage in water. Strain and save the purple juice, which will turn pink in an acidic solution. Add 1/4 cup (85 g) tap water to 1/4 cup of the juice and look for a color change. Next, add 1/4 cup carbonated water to 1/4 cup of cabbage juice. It will turn slightly pink, because when carbon dioxide dissolves in water, carbonic acid is produced.

Jodie Darquea Arteaga

BYCATCH REDUCTION

A DIVERSE LANDSCAPE

Jodie Darquea Arteaga was born in Ecuador, a country on South America's west coast that sits on both sides of the equator and encompasses the Andean highlands, the Amazon jungle, and the Galápagos Islands. When she was sixteen, Jodie went to study biology at the State University, where her research involved learning to breed and harvest a species of fish called almaco jack.

FROM SHRIMP FARMING TO RESEARCH

While Jodie dreamed of getting a master's degree in marine biology, it was difficult for students who were not from wealthy families to advance in the education system, so she went to work for the shrimp industry. However, she was more interested in conservation than farming shrimp and went to work for Equilibrio Azul, a nonprofit research and conservation organization.

ON-BOARD OBSERVATION

Jodie's new position involved accompanying Ecuadorian fishermen on their boats to observe bycatch in small fishing fleets. Bycatch occurs when nontarget species such as dolphins, birds, and turtles get tangled up in fishing lines and nets. These accidental entanglements often result in the death of the bycatch. Jodie monitored hundreds of these fishing trips, documenting the species and bycatch conditions.

SCRIPPS INSTITUTE OF OCEANOGRAPHY

Jodie also fought hard to protect endangered birds like the waved albatross and worked for the government as chief protector of a marine-protected area called the Pelado. Then, to continue her education, she took two years of intensive English classes and applied to graduate school. Her dream came true when she was accepted to the master's program at the University of California's Scripps Institution of Oceanography. The program was difficult, but she persisted, and in 2016 she received her master's degree in advanced studies in marine biodiversity and conservation.

ECUADOR ECOLOGICAL WORLD

Today, Jodie Darquea Arteaga is an associate research professor at Ecuador's State University of the Santa Elena Peninsula, serving as a mentor to other young biologists. She is part of the team at Ecuador Ecological World, a nonprofit organization dedicated to conservation and the sustainable management of coastal marine resources for the benefit of coastal communities.

SAVING SEA TURTLES

Before Jodie's work began, there was no documentation of bycatch in Ecuador. Today, alongside other partners, she and her team continue to work with local fishermen. Jodie's latest project provides LEDs (very bright lights) to fishermen to reduce the number of sea turtles that get tangled in their nets. Jodie, who has always worked hard to achieve her dreams, now has a daughter named Maya to accompany her on her expeditions.

IN TODAY'S WORLD

Sustainable fishing reduces bycatch to protect ecosystems and keep fish populations healthy. When buying seafood, look for the blue Marine Stewardship Council (MSC) logo. The MSC is an international organization whose mission is to end overfishing and ensure seafood is caught sustainably.

TRY THIS AT HOME!

Fill a bowl with water. Add a variety of plastic fish, or small fruit like cranberries, blueberries, or grapes, to the water. Using an aquarium net or a sieved kitchen spoon, try to scoop up one kind of fish or fruit without catching any others.

ABOUT THE AUTHOR

Liz Heinecke has loved science since she was old enough to inspect her first butterfly. After working in molecular biology research for ten years and getting her master's degree, Liz left the lab to kick off a new chapter in her life. Soon she found herself sharing her love of science with her three kids, journaling their science adventures on KitchenPantryScientist.com. Her desire to spread her enthusiasm for science to others soon led to regular TV appearances, speaking engagements, and her books: *Kitchen Science Lab for Kids* (Quarry Books), *Outdoor Science Lab for Kids* (Quarry Books), *STEAM Lab for Kids* (Quarry Books), *Star Wars Maker Lab* (DK), *Kitchen Science Lab for Kids: Edible Edition* (Quarry Books), *The Kitchen Pantry Scientist: Chemistry for Kids* (Quarry Books), *The Kitchen Pantry Scientist: Biology for Kids* (Quarry Books), *Sheet Pan Science* (Quarry Books), *The Padawan Cookbook* (Insight Editions), *The Kitchen Pantry Scientist: Ecology* (Quarry Books), and *RADIANT: The Dancer, The Scientist, and a Friendship Forged in Light*, an adult nonfiction narrative about Marie Curie and Loie Fuller (Grand Central Publishing). Most days, you'll find Liz at home in Minnesota. She graduated from Luther College with a BA in art and received her master's degree in bacteriology from the University of Wisconsin, Madison.

ABOUT THE ILLUSTRATOR

Kelly Anne Dalton is an artist, illustrator, and storyteller living in the wild mountains of Montana. Her elegant and enchanting work can be found on everything from board books to middle grade novel covers, home decor and gift products, and stationery lines. When not drawing, daydreaming, and creating new stories and characters, Kelly Anne can be found trail running in the forests near her home.

ACKNOWLEDGMENTS

I'm extremely grateful to each of the helpful, talented people who collaborated to put this book together: my long-time acquiring editor Jonathan Simcosky; art designer Heather Godin, editorial project manager Gabrielle Bethancourt-Hughes; editorial intern Gabrielle Cruz, design intern Meigan English, marketing managers Angela Corpus and Mel Schuit; and the entire design and editing team. I'm lucky to have such a skilled, supportive group to work with. Thank you to Kelly Anne Dalton, who brings the scientists in these pages to life so beautifully, and to my literary agent Victoria Wells Arms. Special thanks to each of the scientists who took time from their busy schedules to answer my questions and tell me stories from the field. And a shout-out to my family and friends, who keep me laughing and learning every day.

ALSO AVAILABLE

ISBN: 978-1-57715-674-1

ISBN: 978-1-57715-675-8

ISBN: 978-1-57715-673-4